FREAKY TRUE SCIENCE

FREAKY STORIES ABOUT ELECTRICITY

MAPLE ELEMENTARY
SCHOOL LIBRARY
CHARDON OHIO

BY AMY HAYES

Gareth Stevens
PUBLISHING

Please visit our website, www.garethstevens.com. For a free color catalog of all our high-quality books, call toll free 1-800-542-2595 or fax 1-877-542-2596.

Library of Congress Cataloging-in-Publication Data

Names: Hayes, Amy, author.
Title: Freaky stories about electricity / Amy Hayes.
Description: New York : Gareth Stevens Publishing, [2017] | Series: Freaky true science | Includes bibliographical references and index.
Identifiers: LCCN 2016006958 | ISBN 9781482448122 (pbk.) | ISBN 9781482448153 (library bound) | ISBN 9781482448146 (6 pack)
Subjects: LCSH: Electrical engineering–History–Juvenile literature. | Electricity–Juvenile literature.
Classification: LCC TK148 .H39 2017 | DDC 537–dc23
LC record available at http://lccn.loc.gov/2016006958

First Edition

Published in 2017 by
Gareth Stevens Publishing
111 East 14th Street, Suite 349
New York, NY 10003

Copyright © 2017 Gareth Stevens Publishing

Designer: Sarah Liddell
Editor: Ryan Nagelhout

Photo credits: Cover, p. 1 (lightning background used throughout book) Noiro/Shutterstock.com; cover, p. 1 (kite and key) Fer Gregory/Shutterstock.com; cover, p. 1 (plug) Suppakij1017/Shutterstock.com; cover, p. 1 (eel used throughout book) Hein Nouwens/Shutterstock.com; pp. 5, 7, 9, 11, 13, 15, 17, 19, 21, 23, 25, 27, 29 (hand used throughout) Helena Ohman/Shutterstock.com; pp. 5, 7, 9, 11, 13, 15, 17, 19, 21, 23, 25, 27, 29 (texture throughout) Alex Gontar/Shutterstock.com; p. 5 KTSDESIGN/SCIENCE PHOTO LIBRARY/Science Photo Library/Getty Images; p. 7 Sipley/ClassicStock/Archive Photos/Getty Images; p. 8 yevgeniy11/Shutterstock.com; p. 9 Nature/UIG/Universal Images Group/Getty Images; p. 11 Gail Johnson/Moment/Getty Images; p. 13 Martin Leigh/Oxford Scientific/Getty Images; p. 15 loraks/Shutterstock.com; p. 17 Jubobroff/Wikimedia Commons; p. 19 (pacemaker) Science Photo Library/Getty Images; p. 19 (EKG) Tewan Banditrukkanka/Shutterstock.com; p. 21 (electric eel) George Grail/National Geographic/Getty Images; p. 21 (platypus) Robin Smith/Photographer's Choice/Getty Images; p. 23 Ted Kinsman/Science Source/Getty Images; p. 25 incredible Arctic/Shutterstock.com; p. 27 (Tesla) Adam Cuerden/Wikimedia Commons; p. 27 (Edison) Mvuijlst/Wikimedia Commons; p. 29 ullstein bild/Contributor/ullstein bild/Getty Images.

All rights reserved. No part of this book may be reproduced in any form without permission in writing from the publisher, except by a reviewer.

Printed in the United States of America

CPSIA compliance information: Batch #CS16GS: For further information contact Gareth Stevens, New York, New York at 1-800-542-2595.

CONTENTS

What Is Electricity? 4

More Than Just a Magic Trick 6

Thunder, Lightning! 8

The Forever Storm 10

Static Cling 12

Painting with Electricity 14

Conductive Current 16

A Spark in the Heart 18

Electric Animals: Eels 20

Making Light 22

The Aurora Borealis 24

The War of the Currents 26

Tesla's Freaky Inventions 28

Glossary 30

For More Information 31

Index 32

Words in the glossary appear in **bold** type
the first time they are used in the text.

WHAT IS ELECTRICITY?

Ever felt a static charge? Maybe you've rubbed your feet across a carpet and touched something and felt a jolt, or pulled off socks stuck to a shirt when you've taken them out of the laundry. Maybe you've even seen a spark after you pulled them apart. These shocking events are caused by electricity!

Electricity is all around us. It's part of the air we breathe, the charge in our phone, even the beat of our hearts. But electricity isn't a thing you can touch. Electricity is created when electrons, tiny parts of atoms, flow in a particular way. What does this mean? Electrons repel each other. And when these charged electrons get together, things can get pretty freaky.

FREAKY FACTS!

Atoms that have fewer electrons than protons are positively charged. Atoms that have more electrons than protons are negatively charged.

THERE'S ACTUALLY LOTS OF SPACE BETWEEN THE DIFFERENT PARTS OF ATOMS. IF A HYDROGEN ATOM WERE THE SIZE OF EARTH, THE PROTON AT THE CENTER WOULD BE JUST 600 FEET (200 M) ACROSS!

NEUTRON

ELECTRON

PROTON

ATOMS AND CHARGES

An atom is the smallest form of matter that has all the properties of the matter it makes up. Atoms are made up of electrons, protons, and neutrons. Every electron has a negative charge. Every proton has a positive charge. Neutrons are neutral, meaning they have no charge. Every atom that's neutral has the same number of protons and electrons. However, electrons can zoom off an atom if enough energy is applied. They can also move to other atoms. This is important when thinking about charge.

MORE THAN JUST A MAGIC TRICK

People knew that electricity existed long before they understood how it works. In the past, magicians used electricity as a type of magic trick to shock people—literally. No one knew that it was a powerful force that would change the world.

Benjamin Franklin made the connection between these magic tricks and lightning. He believed they were the same thing and wrote it up in a book called *Experiments and Observations on Electricity*. To prove that he was right, Franklin tested his ideas many times. He invented the first lightning rod and learned how to harness the power of lightning to ring bells he placed through his house. We've come a long way from then, but electricity still has the power to amaze.

FREAKY FACTS!

One of Franklin's favorite inventions was the lightning rod. It's a metal pole put on top of a building. It attracts lightning and safely conducts the electricity to the ground without harming the building or people within it.

A KEY AND A KITE

There's a famous story about Franklin's electricity experiments. According to legend, he knew he needed to attract a lightning **bolt** to convince people that lightning was electricity. He built a kite out of a silk handkerchief and two strips of wood, with a wire and a key attached to the end. The story claims he flew it during a storm, and lightning struck the kite and charged the key. When he examined the key, it's said to have given him a spark.

JOSEPH PRIESTLY PUBLISHED AN ACCOUNT OF THE KITE EXPERIMENT 15 YEARS AFTER IT WAS SUPPOSED TO HAVE HAPPENED. BUT EVEN TODAY, WE AREN'T SURE WHETHER THIS FREAKY EXPERIMENT REALLY HAPPENED.

THUNDER, LIGHTNING!

The world lights up in a flash of white light. An instant later, a great thunderclap crashes overhead, rattling the windows. Looking out your window, another jagged beam of hot white light hits the ground, and thunder booms once again. It's a huge thunderstorm!

Clouds are made of millions and millions of atoms of frozen water—ice. These atoms bump into each other, and the electrons that are a part of the water atoms start to form a charge as they attract and push away from each other. The electrons move to the bottom of the cloud. The earth has a positive charge, and opposites attract. The cloud builds up a charge until the electrons fly toward the ground in a bolt of lightning!

FORKED LIGHTNING

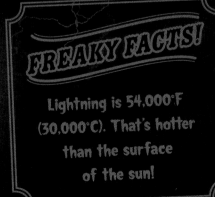

FREAKY FACTS!

Lightning is 54,000°F (30,000°C). That's hotter than the surface of the sun!

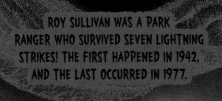

ROY SULLIVAN WAS A PARK RANGER WHO SURVIVED SEVEN LIGHTNING STRIKES! THE FIRST HAPPENED IN 1942, AND THE LAST OCCURRED IN 1977.

INTRACLOUD LIGHTNING

TYPES OF LIGHTNING

There are many different types of lightning. Some basic types are:

intracloud lightning: lightning that looks like a bright flash of flickering light. This occurs when there are positive and negative charges in the same cloud. It's the most common form of lightning.

forked lightning: lightning that looks like a jagged line that hits the ground. Forked lightning can be very dramatic. A bolt of forked lightning can touch the ground over 10 miles (16 km) away!

THE FOREVER STORM

There's a common saying that "lightning never strikes the same place twice." But if you've ever been to Lake Maracaibo in Venezuela, you know that couldn't be more wrong. The lake earned a place in *The Guinness Book of World Records* for most lightning in an area with 250 lightning strikes per square kilometer (0.4 sq mi). The lightning lights up the sky most of the year, and in fall, lightning strikes 28 times a minute!

The **phenomenon** has been called the Beacon of Maracaibo and also Catatumbo lightning. It's made many scientists scratch their head trying to figure out what causes it to happen. While the title "**everlasting** storm" isn't really true, the peak of a storm's bright flashes can certainly seem endless!

FREAKY FACTS!

The area where Lake Maracaibo and the Catatumbo River meet in Venezuela averages 260 storm days per year. That means lightning strikes more than 70 percent of the year!

YOU CAN WATCH LIGHTNING STRIKE ONLINE IN REAL TIME ON THE WORLD WIDE LIGHTNING LOCATION NETWORK.

CATATUMBO LIGHTNING

ELECTRIC PLACES

Even though this stunning display is one of the most famous in the world, the Beacon of Maracaibo isn't the most electric place on Earth. The 158-flash-per-square-kilometer village of Kifuka in the Democratic Republic of Congo held that honor until 2014. Then, NASA (National Aeronautics and Space Administration) discovered that the Brahmaputra Valley in eastern India has the highest lightning flash rate in the world during the months of April and May.

STATIC CLING

Remember the socks and the shirt that stuck together and sparked when you pulled them apart? That's an example of static cling, which is caused by static electricity. Another common example is taking a balloon and rubbing it all over the top of your head until your hair stands up.

Static cling is a fun way to experiment with electricity at home, but what's really happening? Remember that electrons are negatively charged. They zoom around the nuclei, or the middle, of the atoms. Protons exist in these nuclei, and they're positively charged. Rubbing two things against each other causes atoms—and their electrons—to make contact. Electrons can then move from one atom to another. The negatively charged electrons on one surface cling to the positively charged atoms on another, and you get static cling!

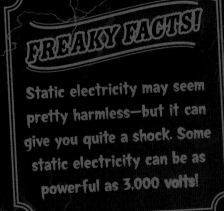

FREAKY FACTS!

Static electricity may seem pretty harmless—but it can give you quite a shock. Some static electricity can be as powerful as 3,000 volts!

FRICTION? I THINK NOT...

Static electricity was first discovered by the ancient Greeks. They believed that static electricity came from rubbing things together. This isn't exactly true. While **friction** may cause static electricity, it isn't the same thing. The reason why rubbing things together creates static electricity is because electrons leave one surface and move to another. But you need specific materials to make static electricity—like woolly socks rubbing on a thick carpet!

WHEN THE WEATHER IS DRY, STATIC ELECTRICITY IS EASY TO CREATE. JUST TRY NOT TO GET SHOCKED!

PAINTING WITH ELECTRICITY

Did you know that people paint with electricity? It's true! People have learned how to use static electricity to make some pretty amazing things! Painting with static electricity is called electrostatic induction. But it isn't something an artist would use to paint on an easel. Instead, carmakers use it when they coat a car with paint.

Electrostatic induction helps paint stick to a car's metal body. Spray guns shoot out paint with strong bursts of air. At the tip of the spray gun is an electrode, which is an element that releases or collects electrons as part of a circuit. The electrode charges the particles of spray paint when they leave the gun. When the negatively charged paint molecules fly out of the spray gun, they cling to the closest object and then dry.

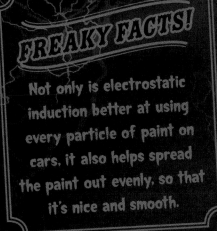

FREAKY FACTS!

Not only is electrostatic induction better at using every particle of paint on cars, it also helps spread the paint out evenly, so that it's nice and smooth.

BECAUSE EVERY MOLECULE OF PAINT IS USED, ELECTROSTATIC INDUCTION IS MUCH MORE **EFFICIENT** THAN OTHER KINDS OF PAINTING

BEYOND PAINT

Electrostatic induction is a pretty cool way to spray-paint. However, there are other ways to use static electricity. Some common machines that use electrostatic induction include laser printers and **photocopiers**. These machines have copying plates that have positive charges and ink that is negatively charged. When the two are put together, the ink sticks to the positively charged plate exactly where the type should go. Next time you print something out, you might be using electrostatic induction.

CONDUCTIVE CURRENT

Different kinds of matter react to electricity differently. Two basic kinds of matter are called conductors and insulators. Conductors are good at allowing electricity to move freely. They're made out of materials that let electrons flow. Some good conductors of electricity are metals such as copper, aluminum, and gold. However, there are other materials that are conductors we don't normally think about, such as water, trees, and people. Remember: people are at least 60 percent water!

While conductors let electrons flow easily from molecule to molecule, insulators make it more difficult for electrons to move. Insulators such as rubber and plastic surround electric wires, making them safe to pick up, even when they're plugged in. Other insulators include glass and porcelain.

FREAKY FACTS!

The first-ever-discovered superconductor was the metal mercury. It's liquid at room temperature, but needs to be much colder to be superconductive!

SUPERCONDUCTORS

A superconductor is matter that conducts electricity without any resistance at all once it reaches a certain temperature. Most of these materials need to be very cold—at least −220°F (−140°C) at normal pressure. Scientists have tried to find superconductors that don't need to be cooled to extreme temperatures. In 2015, they found a superconductor at −94°F (−70°C), a temperature found naturally on Earth! Scientists hope one day they can find a material that achieves superconductivity at room temperature.

SOME SUPERCONDUCTORS CAN FLOAT IF THEY'RE HELD IN PLACE BY MAGNETS!

A SPARK IN THE HEART

We know that humans are good conductors of electricity—that's why it's so dangerous to be outside in a storm. But did you know that every living person is filled with electric impulses? Electricity is an essential part of the human body. It's what causes the muscles in our hearts to contract. Without electricity, we wouldn't be able to pump blood through our bodies!

Our heartbeat is involuntary, which means it beats without us thinking about it. But when someone is having trouble with their heartbeat, they're given an electrocardiogram, or EKG for short. This is taken by an electrocardiograph, a machine that attaches special electrodes to different parts of the body and measures the electrical impulse sent from the heart. When the electricity in a person's body contracts the heart, the electrodes sense it.

FREAKY FACTS!

The number of electrical impulses, or heartbeats, a person has depends on whether they're calm or excited, active or sitting still. The average heartbeat of a healthy adult is between 60 and 100 beats every minute.

IF SOMEONE HAS AN UNUSUAL HEARTBEAT, THEY'RE OFTEN GIVEN A PACEMAKER. IT'S A SMALL DEVICE PLACED UNDER THE SKIN NEAR THE HEART THAT HELPS KEEP A PERSON'S HEART BEATING AT THE RIGHT PACE.

EKG

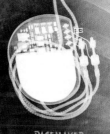

PACEMAKER

HOW TO READ AN EKG

When an electrical impulse flows through the body, the electrocardiograph shows this as a blip on its screen. When a person is healthy, these blips are always the same distance apart, just as a heart beats in a regular pattern. When a person is unhealthy, there might be a pause in between blips or some blips that are too close together. Doctors use EKGs to make sure people are healthy and figure out how to help people who aren't.

ELECTRIC ANIMALS: EELS

Since we know that heartbeats are caused by electricity, it makes sense to think that most animals have the same kind of electricity as people. However, some definitely have more ability to shock than others! Enter the electric eel.

Electric eels aren't actually eels at all—they're a type of fish. They live in streams and ponds in South America and eat fish, small mammals, birds, and amphibians such as frogs. But that's not the freaky part. These eels have the ability to electrify the water around them! They actually shock their prey with electricity! Each electric eel has 6,000 special cells called electrocytes. These electrocytes store power somewhat like batteries do. If an eel sees an animal it wants to attack or feels scared by a predator, the electrocytes send out a powerful shock.

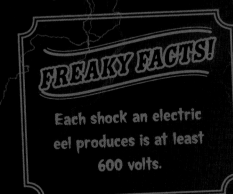

FREAKY FACTS!

Each shock an electric eel produces is at least 600 volts.

ELECTRIC EELS ARE USUALLY 6 TO 8 FEET (1.8 TO 2.4 M) LONG AND WEIGH AROUND 44 POUNDS (20 KG). LUCKILY, THEY USUALLY LEAVE HUMANS ALONE.

ELECTRIC EEL

PLATYPUS

THE PLATYPUS

The duck-billed platypus has a bill like a bird, fur like a mammal, and lays eggs! But did you know duck-billed platypuses are electric creatures? The bill of a platypus is filled with cells called electroreceptors. Its electroreceptors sense tiny currents caused by the nerves and muscles of other animals around it. These electroreceptors help the duck-billed platypus hunt in the dark!

MAKING LIGHT

Triboluminescence is a very long word for a very weird phenomenon. "Tribo" means "friction," and "luminescence" means "to give off light." So triboluminescence is friction that gives off light. This can happen, for example, when two pieces of a mineral called quartz are scraped across each other. The quartz pieces sometimes create a spark of light. This is triboluminescence.

Scientists still don't know exactly why triboluminescence happens. Some think the charge created by the friction makes gas trapped in the minerals light up. They do know that it doesn't work with just any old rocks, though. Quartz is a very common rock that can create this flash of light, but other minerals such as sphalerite, fluorite, calcite, muscovite, and some opals can also create this effect.

FREAKY FACTS!

Think triboluminescence is hard to spell? It's also called fractoluminescence, or mechanoluminescence.

ABOUT HALF OF THE MANY DIFFERENT TYPES OF CRYSTALS IN THE WORLD CAN SPARK WITH THE RIGHT AMOUNT OF FRICTION.

HAMMER

MINT

SPARKS IN THE CHEEKS

One famous way people experiment with triboluminescence is by eating mint candy. A person biting down on a certain kind of breath mint creates friction. This friction actually creates a charge in the candy, as the pieces separate into positive and negative charges when they break into pieces. Then these charges hit the air and combine with a gas called nitrogen to give off a bit of blue light. It might be hard to see unless the lights are off, though.

THE AURORA BOREALIS

If you ever get to travel way up north, try to spot the beautiful electric light show that happens in the night sky. The aurora borealis, sometimes called the northern lights, is a stunning wonder of the natural world. Colors dance across the sky as though someone had painted them there. But the aurora, however, colors the night sky green, blue, and red for a very scientific reason. It's electricity!

The sun doesn't just send light and heat to Earth, but also a large number of charged electrons. These streams of electrons are part of **solar wind**. When this wind hits Earth's atmosphere, the electrons flow towards the North and South Poles. Then they interact with elements in the atmosphere and light up in the night sky high above the ground.

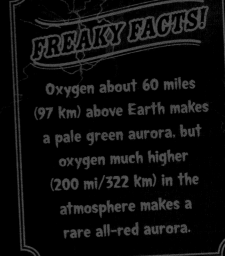

FREAKY FACTS!

Oxygen about 60 miles (97 km) above Earth makes a pale green aurora, but oxygen much higher (200 mi/322 km) in the atmosphere makes a rare all-red aurora.

WHILE THE AURORA AT THE NORTH POLE IS CALLED THE AURORA BOREALIS, THE ONE AT THE SOUTH POLE IS CALLED THE AURORA AUSTRALIS.

WHAT COLOR IS ELECTRICITY?

The aurora borealis is filled with different-colored lights—some are a deep red and purple, some are different shades of green or yellow. This is because atoms emit different-colored lights when exposed to energy like electricity. In fact, each element has its own specific color spectrum. For example, when electrons and oxygen combine, they create green or yellow, while when electrons collide with nitrogen, they give off a red, purple, or blue.

THE WAR OF THE CURRENTS

Figuring out the best way to move and use electricity led to a public battle between two famous scientists: Thomas Edison and Nikola Tesla. Tesla actually worked for Edison's company before the two had a falling out. Tesla then joined George Westinghouse's company and worked on alternate current (AC), a rival to Edison's direct current (DC). While direct current only allowed for the flow of energy in one direction, alternating current let energy flow back and forth in a loop.

Alternating current won out as the more useful current, but not before Edison launched a **propaganda** war to hurt AC's reputation. He called AC dangerous and even **electrocuted** dogs and horses with it to show its danger. The fight between the two inventors came to be known as the War of the Currents.

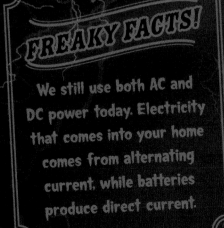

FREAKY FACTS!

We still use both AC and DC power today. Electricity that comes into your home comes from alternating current, while batteries produce direct current.

HERE'S WHY TESLA'S AC POWER WON OUT OVER EDISON'S DC POWER.

POWER ON THE MOVE

	ALTERNATING CURRENT	DIRECT CURRENT
INVENTOR	NIKOLA TESLA	THOMAS EDISON
DISTANCE ELECTRICITY CAN TRAVEL	HUNDREDS OF MILES	ABOUT A MILE
ABILITY TO CHANGE VOLTAGES	EASY	DIFFICULT
CURRENT FLOWS	BOTH DIRECTIONS	ONE DIRECTION

NIKOLA TESLA   THOMAS EDISON

WHO INVENTED THE LIGHT BULB?

This seems like a simple question, but there's no easy answer. In 1800, Alessandro Volta made a battery that could light up a copper wire. Then, in 1802, Humphrey Davy created the arc lamp. In 1850, Joseph Swan invented the first true light bulb, but it never made it to market. Several others worked on different lights that were expensive. Edison didn't really invent the light bulb—his claim to fame was actually a bulb that people could afford!

TESLA'S FREAKY INVENTIONS

Tesla's AC power system was just one of his many electric inventions. He designed and built an alternating current motor in 1883 and later used it to build a **hydroelectric** power plant in Niagara Falls, New York—the first ever built. He also designed "Tesla coils," which are large **transformers** that transmit electricity wirelessly and shoot lightning out of them! Tesla coils create amazing light shows, but are mostly used for entertainment today.

Tesla also experimented with sending energy over long distances wirelessly. He built a tower at a place called Wardenclyffe in Shoreham, Long Island, that was supposed to broadcast electricity and forms of communication. A wealthy banker named J. P. Morgan, however, pulled his funding, and the project was stopped. Maybe some ideas are just too freaky for some people!

FREAKY FACTS!

Tesla also created an electric boat he could control remotely. He even developed many of the communications tools needed for the radio, despite the fact that Guglielmo Marconi is usually credited for its invention.

DEATH RAY OR PEACE BEAM?

One of Tesla's freakiest inventions was a ray gun the inventor thought of when he was 78. He thought the weapon, which shot a beam of particles through the air, could destroy weapons, planes, and even troops. Tesla thought his "peace beam" could stop war because no one would be able to attack one another. However, people were afraid the beam could actually be used for war. He was never able to complete the weapon.

ONE OF TESLA'S MANY UNUSED IDEAS WAS TO HEAT THE UPPER ATMOSPHERE OF EARTH UP, CREATING MAN-MADE AURORAS!

GLOSSARY

bolt: a strike of lightning

efficient: the least wasteful means of doing a task

electrocute: to kill using electricity

everlasting: never ending

friction: resistance to the relative motion of one body moving over another with which it is in contact

hydroelectric: having to do with the production of electricity by waterpower

phenomenon: a rare fact or event

photocopier: a machine that makes an exact copy of a printing or drawing

propaganda: the spread of ideas to hurt or help an institution, cause, or person

solar wind: a flow of charged particles that is sent from the sun out into space

transformer: a device that converts variations of current in a primary circuit into variations of voltage and current in a secondary circuit

volt: a unit used to measure electricity

FOR MORE INFORMATION

BOOKS

Graham, Ian. *You Wouldn't Want to Live Without Electricity!* New York, NY: Franklin Watts, 2015.

Parker, Steve. *Electricity.* New York, NY: DK, 2013.

WEBSITES

Electricity Facts
sciencekids.co.nz/sciencefacts/electricity.html
Check out a bunch of cool facts about electricity.

Fun Facts About Electricity
artisanelectric.com/fun-facts-about-electricity/
Find out more fun facts about electricity here.

World Wide Lightning Location Network
wwlln.net/new/map/
Track every bolt of lightning around the globe.

Publisher's note to educators and parents: Our editors have carefully reviewed these websites to ensure that they are suitable for students. Many websites change frequently, however, and we cannot guarantee that a site's future contents will continue to meet our high standards of quality and educational value. Be advised that students should be closely supervised whenever they access the Internet.

INDEX

atoms 4, 5, 8, 12
aurora borealis 24, 25
Beacon of Maracaibo 10, 11
Brahmaputra Valley 11
Catatumbo lightning 10
charges 5, 8, 9, 12, 14, 23
conductors 16, 18
duck-billed platypus 21
Edison, Thomas 26, 27
EKG 18, 19
electric eels 20, 21
electrocardiograph 18, 19
electrocytes 20
electrons 4, 5, 8, 12, 13, 14, 16, 24, 25
electroreceptors 21
electrostatic induction 14, 15
Franklin, Benjamin 6, 7
friction 13, 22, 23
heartbeat 18, 19
insulators 16
Kifuka 11
Lake Maracaibo 10
lightning 6, 7, 8, 9, 10, 11
protons 4, 5, 12
ray gun 29
solar wind 24
static cling 12
static electricity 12, 13, 14, 15
superconductors 16, 17
Tesla coils 28
Tesla, Nikola 26, 27, 28, 29
triboluminescence 22, 23
volts 12, 20
War of the Currents 26